BEYOND THE ALGORITHM

HUMAN-CENTERED AI IN COUNSELING

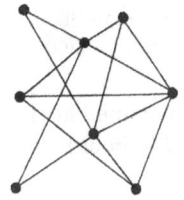

DR. NARINEH MAKIJAN

Tulsa, Oklahoma | Guadalajara, Mexico

Beyond the Algorithm: Human-Centered AI in Counseling
©2025 by Dr. Narineh Makijan

Published by Grafo Education
An imprint of Grafo House Publishing
Tulsa, Oklahoma | Guadalajara, Mexico

ISBN   978-1-963127-30-0 (paperback)
        978-1-963127-31-7 (e-book)

All rights reserved. No part of this publication may be reproduced in any form or by any electronic or mechanical means, including information storage and retrieval systems, without permission in writing from the publisher, except by a reviewer who may quote brief passages in a review. The views and opinions expressed herein are solely those of the authors and do not necessarily represent the views and opinions of the publisher.

Cover design by ChatGPT with human editing and design.

Siri is a trademark of Apple Inc., registered in the U.S. and other countries and regions.

Printed in the United States of America
28 27 26 25   1 2 3 4

To my children, Leon, Liam, and Sophia—your love, strength, and curiosity are my greatest inspiration. You remind me each day why this work matters.

And to every student, counselor, and educator committed to leading with empathy in an ever-evolving world, this book is for you.

Table of Contents

Introduction   1

1 Why This Book, and Why Now?   5

2 Understanding AI in Counseling Contexts   15

3 The Heart of Counseling is Human Connection   27

4 Ethical Imperatives in the Use of AI   35

5 AI Through a Cultural and Equity Lens   49

6 Collaboration, Not Replacement   61

7 Developing AI Literacy for Counselors   71

8 Real-World Applications   81

9 Student Voice and Well-Being in the Age of AI   95

10 Looking Forward   105

Appendices

1 Student Voice Engagement Toolkit — 115

2 AI Implementation Planning Template — 121

3 Counselor's AI Evaluation Checklist — 127

4 Student Consent & Transparency Template for AI Tools — 129

5 Equity Review Questions for School Teams — 131

6 AI Tool Pilot Planning Guide — 133

Endnotes — 135

Introduction

"Hey Siri, when is my counseling session?"

Five years ago, this sentence might have sounded absurd. But today, students are being directed to AI-powered bots for everything from college applications to emotional check-ins. For over two decades, I've worked alongside students facing anxiety, food insecurity, academic pressure, identity exploration, and hope for a better future. What I've learned is this: nothing replaces a human who listens fully, without judgment.

Yet, we can't ignore the possibilities. AI tools can provide support when used wisely. They can surface early warning signs, streamline tasks, or offer 24/7 check-ins. But they can also mislabel, misinterpret, and marginalize. When these systems are trained on biased data or applied without cultural context, they can cause harm.

As artificial intelligence rapidly transforms education, counselors are increasingly expected to navigate AI tools—often without adequate guidance or resources. Currently, there is a lack of comprehensive, practical guidance for K16 counselors on how to ethically and equitably integrate AI into student support services. This is especially critical for vulnerable, underrepresented, and neurodiverse student populations, who may be disproportionately affected by AI tools that lack cultural sensitivity or fail to account for diverse learning needs.

This book bridges that gap by offering a practical, human-centered framework for integrating AI ethically and equitably into K16 counseling. With extensive experience in education, leadership, and counseling, I offer tools and actionable strategies to help counselors ensure AI enhances, rather than replaces, the relational core of student support.

I'm sure you don't need me to tell you that the challenges and realities facing educational counselors in the age of AI are woefully under-researched. Current statistics are often unavailable, and individual schools and districts may have widely differing experiences. This book is intended to be a guide toward practical action for counselors in the field rather than a detailed analysis of the state of AI implementation—which will always be a moving target, anyway. Some statistics are from the early days of ChatGPT and other generative AI models, and therefore are not as current as I would like; but they are enough to give a sense of the challenges facing counselors in educational settings.

With the goal of being as practical and readable as possible, chapters are intentionally concise and action-driven. I have included observations and steps that have served me and those I work with over the years, trusting readers to find unique, tangible ways to apply them in their respective contexts.

The book addresses the pressing need for comprehensive guidance on ethically integrating AI into student support services, particularly for vulnerable, underrepresented, and neurodiverse student populations. It emphasizes the importance of cultural responsiveness, equity-driven evaluation, and the protection of student agency in the age of AI. By centering student voices and well-being, it provides a roadmap for counselors to lead in the ethical adoption of AI technologies in educational settings.

This book—and the task of the counselor—is about balance. It's about empowering counselors to become not just passive users, but critical stewards of AI in their schools. It's about making sure that equity and empathy lead every conversation we have about innovation.

Our students deserve nothing less.

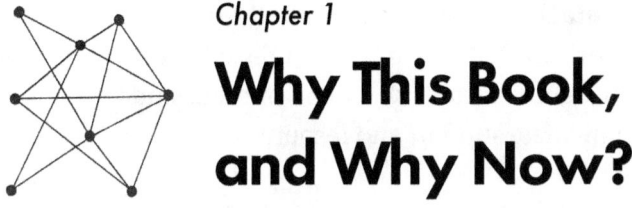

Chapter 1

Why This Book, and Why Now?

Education is undergoing a seismic transformation—one that doesn't always make headlines, but is happening in classrooms, counseling offices, and digital dashboards across the country. Artificial Intelligence (AI) has arrived not as some distant threat or novelty, but as a reality that is rapidly reshaping how institutions function and how students are supported.

In recent years, AI has become deeply embedded in education systems. In fact, according to HolonIQ 2025 Global Education Outlook, the global education market is projected to reach nearly $10 trillion by 2030, with AI moving from experimental phases to practical implementation across various educational applications.[1] In the United States, educational institutions are increasingly adopting AI tools for:

- **Automated grading and content generation**
  AI-powered platforms like Magic School AI and Brisk Teaching are streamlining educator workloads by automating grading processes and generating educational content.

- **Mental health support via chatbots**
  AI-driven chatbots are being utilized to provide mental health check-ins, offering students immediate support and resources.

- **Career and college advising platforms**
  Platforms such as SkillUp, PraktikaAI, Naviance, Xello, and Major Clarity are enhancing career and college advising by providing personalized guidance to students.

- **Early alert systems for dropout prevention**
  Institutions are leveraging AI to analyze student data and identify those at risk of dropping out, enabling timely interventions.

Tools like Khanmigo, Sown to Grow, and YouScience Brightpath are being deployed in both K12 and postsecondary environments, promising efficiency, personalization, and early intervention. Yet, amid this digital evolution, a critical stakeholder group has been largely left out of the conversation: counselors.

These are the professionals uniquely trained to safeguard student mental health, support academic and career development, and uphold equity and inclusion. Yet they are now expected to make sense of algorithmic flags, manage AI-generated student profiles, and respond to digital wellness dashboards—often with no formal training, no ethical guidance, and no seat at the decision-making table.

This book is an answer to that gap.

A Personal Imperative

After more than two decades in K16 education, as an administrator, counselor, professor, and regional leader, I've seen firsthand the disconnect between technology adoption and human readiness. I've witnessed students lose trust in systems that mislabel them as "at risk" based on data alone. I've supported counselors struggling to interpret algorithm-generated alerts without understanding the source or context. I've seen AI used as both a bridge to care and a barrier to belonging.

These experiences are not isolated.

A 2025 study found that while 62% of U.S. educators regularly use AI in their professional roles, 69% reported receiving little to no formal training from their schools. This gap between high adoption and limited preparation hinders the effective and responsible use of AI tools. Notably, there remains a significant lack of data specifically capturing counselors' perspectives and usage, leaving their unique needs and experiences largely unaddressed in the broader conversation on AI in education.[2]

Meanwhile, the student-to-counselor ratio remains unacceptably high—385:1 nationwide, far above the 250:1 ratio recommended by the American School Counselor Association (ASCA). In this environment, school districts are increasingly turning to AI as a solution to overloaded counseling systems. Hence, it is critical that counselors take the lead and become stewards of AI implementation, ensuring they receive proper training and are actively involved in shaping how these tools are used to support, not substitute, the human-centered foundation of counseling.[3]

But AI is not neutral. When trained on biased or incomplete data, it can reinforce disparities in discipline, academic tracking, and postsecondary opportunities. When deployed without transparency or consent, it can erode trust and student agency. And when used to replace, rather than augment, human connection, it can do real harm.

This book is not an indictment of technology—it is a call to reclaim the counselor's role in shaping how technology is designed, implemented, and evaluated. It is an invitation to engage with AI critically, ethically, and empathetically.

Why? Because students deserve more than systems that track them. They deserve systems that see them.

The Stakes Are High

Why does this matter right now?

Because AI is rapidly outpacing our school systems' ability to ethically integrate it. Decisions about which tools to use and how to use them are often made by vendors, administrators, or IT departments, without counselor input. These tools are marketed as scalable solutions to mental health needs, career counseling shortages, and learning loss, which are complex, fundamental human problems.

If counselors are not at the table, we risk:

- Reinforcing existing inequities through algorithmic bias
- Automating interventions without relational support
- Eroding confidentiality, consent, and student trust
- Prioritizing efficiency over empathy

In 2023, UNESCO released a global policy guide urging school systems to center ethics, equity, and human rights in the use of AI. Meanwhile, organizations like the ASCA and the American Counseling Association (ACA) have begun releasing position statements; however, practical implementation guidance focused on K16 remains scarce.

What to Expect

Each chapter of this book is designed to bridge the gap between AI's technical possibilities and counseling's human-centered mission. Whether you are a counselor, administrator, EdTech developer, or graduate student in education, this book will help you engage more confidently and critically in the conversation.

You will find:

- Plain-language explanations of key AI concepts including machine learning, Natural Language Processing (NLP), and predictive analytics
- Real-world case studies from schools across the U.S.
- Ethical checklists and reflection tools tailored to counseling practice
- Culturally responsive strategies for working with diverse students
- Policy guidance for advocating at the school, district, institution, or state level

This book offers not just theory, but action rooted in lived experience, professional ethics, and a commitment to student dignity.

The future of counseling is not human only, nor is it AI only. It is human-centered, ethically guided, and AI-informed.

Reflection Questions

1. In what ways is AI already impacting your institution or counseling environment, even if subtly?

2. Have you ever felt unprepared or uninformed when faced with a new tech tool or data dashboard? What emotions did that evoke?

3. How often are counselors consulted during the selection or implementation of EdTech tools in your institution?

4. When have you witnessed a disconnect between technology and student needs? What was the outcome?

5. What concerns do you have about AI in education, particularly regarding ethics, equity, and student well-being?

6. What opportunities do you see for AI to enhance your work as a counselor, if implemented thoughtfully?

7. Do you feel empowered to speak up about tech integration at your institution? Why or why not?

8. How do your own lived experiences influence how you view the role of human connection in counseling?

Action Steps

1. Start a conversation with a colleague, administrator, or tech lead about how AI tools are being selected and used at your site. Ask whether counselors are part of that process.

2. Review your district's current EdTech tools and identify any that use AI or predictive analytics. Make note of how they intersect with student mental health, academic planning, or behavior tracking.

3. Conduct an informal survey among fellow counselors or educators. Have they received any training on AI? Do they feel confident in using it responsibly?

4. Journal your observations over the next week about any interactions—direct or indirect—with AI in your school setting. What did you notice? How did it align (or conflict) with student-centered practice?

5. Download and review any available ethical guidelines from ASCA, ACA, or United Nations Educational, Scientific and Cultural Organization (UNESCO) related to AI use in schools. Highlight sections that resonate with your role.

6. Schedule a PD session or book club around this text or a related topic. Use it to create space for counselors to collectively reflect and build capacity.

7. Advocate for inclusion. Bring up counselor involvement in tech decision-making at your next staff, department, or leadership meeting.

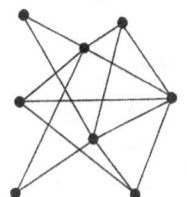

Chapter 2

Understanding AI in Counseling Contexts

At first glance, artificial intelligence and counseling may seem like distant domains. One is grounded in algorithms and automation and the other in empathy and human connection. But in today's rapidly evolving educational landscape, these two worlds are converging with increasing speed. For counselors, gaining a foundational understanding of AI is essential.

The purpose of this chapter is not to turn counselors into coders. Rather, it's to empower them to engage critically with the technologies entering their spaces, ask the right questions, and ensure every AI tool used in schools reflects the values of trust, ethics, and student-centered care.

What Is AI, Really?

Artificial Intelligence (AI) refers to technologies that simulate human intelligence to perform tasks like recognizing speech, interpreting language, identifying patterns, and making decisions.

Key types of AI relevant to counseling include:

- **Machine Learning (ML)**
  Systems that improve their predictions over time based on data inputs (e.g., identifying students at risk of academic failure).

- **Natural Language Processing (NLP)**
  Tools that allow machines to understand and respond to human language (e.g., mental health chatbots, journaling apps).

- **Recommendation Systems**
  Platforms that suggest personalized pathways, resources, or opportunities based on behavioral patterns and user profiles (e.g., career and college guidance tools).

These technologies are often built into tools counselors are already using, from early-alert dashboards to online advising platforms.

AI in K16 Counseling: What's Happening Now

According to a 2025 survey by the EdWeek Research Center, 43% of K–12 teachers reported receiving at least one training session on AI, marking a significant increase from previous years.[4] However, the survey did not provide specific data regarding training for counselors or mental health staff. This lack of targeted information underscores the need for more comprehensive research into AI training and usage among counselors.

Given the increasing reliance on AI tools in educational settings, it's crucial to understand how counselors are being prepared to integrate these technologies into their practice. Without adequate training and data, there's a risk that AI implementation may not effectively support the unique needs of student mental health and counseling services.

Further studies focusing on the perspectives and training of counselors in AI usage are essential to ensure that these professionals are equipped to leverage AI tools responsibly and effectively in their roles.

As of 2025, AI applications in counseling continue to expand, addressing various aspects of student support. Here are some of the most prevalent tools and platforms:

- **Early alert systems platforms** like Starfish, Navigate360, or Panorama Education use AI to assess behavioral patterns, attendance, and grades, flagging students for early intervention.

- **Mental health and emotional check-ins:** AI tools like Kooth, Woebot, and Sentinel use NLP to detect emotional tone and provide responses. Some school districts have adopted mood-check-in surveys that alert counselors based on keywords or stress indicators.

- **Career and college readiness platforms tools** such as Naviance, MajorClarity, and Xello use AI to suggest postsecondary pathways. These systems evaluate interests, behavior, and even social media data in some cases.

- **Scheduling and workflow automation**: AI-based platforms can automate appointment setting, reminder texts, or progress tracking. This reduces administrative burdens and allows counselors to focus more time on direct support.

Benefits and Possibilities

When used intentionally and ethically, AI can enhance, not hinder, counseling. This illustrates the way in which AI will improve:

- **Efficiency**
  Automates routine tasks and frees counselors for student interaction.

- **Proactive support**
  Flags students for support earlier than traditional systems might allow.

- **Scalability**
  Offers support in under-resourced schools where counselor caseloads are overwhelming.

- **Insights**
  Helps identify patterns in student needs or engagement across large populations.

According to McKinsey's 2025 "Superagency in the Workplace" report, the future of AI in professional settings extends beyond efficiency—it lies in its potential to enhance human agency.[5] Emerging uses of conversational and reasoning-based AI are already supporting individuals in areas like counseling, coaching, and creative expression. These tools are not just automating tasks; they are beginning to mirror emotional intelligence, offering real-time support that complements human connection.

For counselors, this signals a pivotal opportunity: by thoughtfully integrating emotionally intelligent AI tools, educators can expand access to wellness supports and reimagine how students engage with guidance services, all without compromising the relational heart of the profession.

Risks and Limitations

Without thoughtful implementation, AI can create unintended harm:

- **Bias and inaccuracy**
  AI systems have been found to reflect and perpetuate existing inequities, especially in communities of color and for students with disabilities.

- **Lack of context**
  AI may miss the nuances of trauma, home life, or identity, which are critical elements in counseling.

- **Privacy risks**
  AI systems store sensitive student data, raising significant privacy concerns. The Center for Democracy & Technology (CDT) highlights the need for transparency and greater parental engagement, emphasizing that schools must clearly communicate how AI and other technologies are used, along with their potential risks.

- **Overreliance**
  Schools may begin to favor efficiency over empathy, using tech in place of trained professionals.

Questions Counselors Should Be Asking

As AI becomes more integrated, counselors must take an active role in evaluating tools. Key questions include:

- Who developed this tool, and were counselors involved?
- What data is being used, and is it culturally/contextually appropriate?
- Has this system been audited for bias?
- Is student data stored securely and ethically?
- Do students and families understand the tool, and can they opt out?
- How are flagged concerns followed up—with humans or with automation?

Questions such as these are ethical imperatives that tie directly to the core values of counseling: safety, confidentiality, advocacy, and student voice.

Building AI Literacy = Counselor Empowerment

You don't need a computer science background to lead in this space. But you do need:

- Basic knowledge of how AI works
- Confidence to ask hard questions

- Awareness of red flags
- Commitment to human dignity in tech use

As ASCA stated in its 2023 statement "The School Counselor and Student Safety with Digital Technology":

> School counselors have a responsibility to promote healthy student development and to protect students from digital technology's potential risks. School counselors consider the ethical and legal considerations of technological applications, including confidentiality concerns, student and community safety concerns, security issues, potential benefits and limitations of communication practices using electronic media, and managing appropriate boundaries with students and stakeholders.[6]

ASCA's position underscores the necessity for counselors to be proactive in understanding and managing the implications of AI in education. By adhering to ethical standards, collaborating with stakeholders, and engaging in continuous learning, counselors can effectively integrate AI tools to enhance student support while safeguarding their well-being. When counselors gain AI literacy, they protect students and position themselves as vital leaders in shaping the future of education.

Resources for Continued Learning

- Common Sense Education AI Toolkit: https://www.commonsense.org/education/lists/classroom-tools-that-use-ai
- Center for Humane Technology: https://www.humanetech.com
- ASCA – Emerging Tech and School Counseling: https://www.schoolcounselor.org
- Google's Teachable Machine: https://teachablemachine.withgoogle.com
- Future of Privacy Forum: https://fpf.org

In the next chapter, we return to the foundation of our profession: the human relationship. As AI grows more capable, counselors must hold firm to the values that no machine can replicate: empathy, cultural attunement, and the healing presence of a caring adult. Because while AI can identify a student in distress, only a counselor can truly see them.

Reflection Questions

1. Have you encountered any AI-powered tools in your current counseling practice? If so, how were you introduced to them?

2. Do you feel confident in your understanding of how AI tools function behind the scenes? Why or why not?

3. How do you currently evaluate whether an educational technology aligns with ethical or student-centered counseling principles?

4. What concerns do you have about how AI might interpret or misinterpret student behaviors or emotional states?

5. Which of the "Questions Counselors Should Be Asking" in this chapter resonates most with you and why?

6. How prepared do you feel to advocate for ethical AI use in conversations with school leaders or vendors?

7. Where do you see opportunities for AI to support your work more effectively without replacing the human connection?

8. Reflecting on your own school or district: Are students and families informed about the technologies being used to track or support them? Should they be more involved in those decisions?

Action Steps

1. Make a list of all digital platforms used at your school that involve student data, advising, behavior tracking, or mental health check-ins. Research whether any of them use AI or predictive analytics.

2. Use one of the "Questions Counselors Should Be Asking" from this chapter in your next staff or department meeting. Invite discussion on how ethical considerations are currently addressed.

3. Identify one AI concept (e.g., NLP, machine learning) that feels unfamiliar, and take thirty minutes to explore it using the recommended resources or other trusted sites.

4. Connect with your IT or EdTech department to ask whether any ethical reviews or bias audits have been conducted on current AI tools in use.

5. Create a brief "AI Awareness Tip Sheet" for fellow counselors or administrators that outlines the benefits, risks, and essential questions from this chapter.

6. Schedule a learning circle or informal training session for your counseling team using the Common Sense Education AI Toolkit or the ASCA Emerging Technologies resources.

7. Initiate a conversation with student leaders or families about how AI is being used at your school and what transparency or consent practices should be in place.

8. Keep a reflection log over the next month noting any interaction with AI in your daily work—whether direct or indirect—and how it impacted your ability to support students.

Chapter 3

The Heart of Counseling Is Human Connection

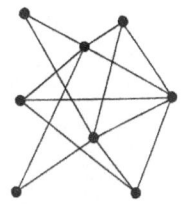

At the center of every meaningful counseling interaction is a relationship built on empathy, trust, understanding, and presence. Counselors do more than deliver information or manage student records. They bear witness to a student's lived experiences, helping them navigate not only academic goals, but also personal growth, identity formation, trauma, and hope.

Technology, no matter how advanced, cannot replicate this. As AI enters schools, we must reaffirm that counseling is not simply a transaction. It is a relational practice rooted in humanity.

What Makes Counseling Human?

Counseling is guided by several foundational principles:

- **Empathy**
  The ability to feel with, not just for, a student

- **Unconditional positive regard**
  Accepting and supporting students without judgment

- **Cultural responsiveness**
  Understanding and respecting diverse backgrounds and identities

- **Ethical integrity**
  Upholding confidentiality, trust, and care

- **Developmental understanding**
  Recognizing that students are growing, changing, and vulnerable

These principles cannot be programmed into a machine. AI can analyze data, simulate responses, and even respond with warmth, but it lacks true emotional attunement. A chatbot might say "I'm sorry you're feeling sad," but it will never offer the grounding presence of a counselor sitting beside a student in distress.

The Danger of Replacing Human Support

As funding tightens and student needs increase, some districts are tempted to lean more heavily on AI tools for emotional and academic support. While these tools can complement counseling efforts, they are never a substitute.

Case in point: A high school rolled out an AI-powered emotional wellness app designed to check in with students each morning. Within weeks, the system flagged a student multiple times for high distress. The app recommended journaling and calming videos, which the student ignored. No counselor follow-up was initiated because the school assumed the tool was sufficient. Two months later, the student dropped out after experiencing prolonged depression without support.

The problem wasn't the tool; it was the absence of a human response. The signals were there, but no one was listening beyond the algorithm.

This is not an isolated scenario. According to the CDC's Youth Risk Behavior Survey (2013–2023), more than 42% of high school students reported feeling persistently sad or hopeless in the past year, a number that has risen steadily over the last decade. Yet only 1 in 5 students who experience mental health challenges receive school-based support.[7]

Holding Space in a Digital Age

Students need more than AI-generated suggestions. They need safe spaces to be seen, heard, and guided. Counselors offer this by:

- Listening without rushing to solve
- Responding with cultural and developmental insight

- Noticing nonverbal cues that data doesn't capture
- Recognizing when silence speaks volumes

The Supporting Minds, Supporting Learners (CCCSE, 2024) report underscores that when students face emotional challenges such as grief, fear, identity confusion, or trauma, they seek human connection, not technology. Nearly half of students (49%) prefer to talk to a friend, partner, or family member, while another 37% prefer a trained mental health provider, highlighting a strong preference for trusted relationships over impersonal support systems. The study also found that 56% of students reported that mental health difficulties hurt their academic performance in the past four weeks, and students with likely depressive or anxiety disorders were nearly twice as likely to experience academic disruptions compared to their peers.[8] These findings affirm that during moments of emotional vulnerability, what students need most is empathy, trust, and relational support, not predictive analytics or automated interventions.

AI can serve as a powerful supportive tool when grounded in human relationships. Imagine the impact of:

- A counselor using AI to monitor early warning signs but following up personally with each student.
- A chatbot offering after-hours support, with counselors reviewing logs to deepen their understanding.

- Scheduling tools freeing up time for more one-on-one sessions.

The key is intentional integration, not substitution.

The Irreplaceable Role of the Counselor

In a world that increasingly values speed and efficiency, counselors remind us of the power of slowing down. The strength of the profession lies not in how quickly we can identify issues, but in how deeply we can respond to them.

While AI can support and enrich our work, it should never define it. As educators, advocates, and healers, counselors have a profound responsibility to create meaningful, human-centered connections that no algorithm can replicate.

In the next chapter, we'll explore the ethical terrain of using AI in counseling, including how to safeguard student rights, mitigate bias, and maintain trust.

Reflection Questions

1. When was the last time you felt that a counseling session had a profound impact on a student? What made that moment meaningful?

2. How do you ensure that students feel truly seen and heard in your practice, especially when time and resources are limited?

3. Have you ever witnessed technology being used in a way that unintentionally diminished human connection? How did you respond?

4. What are the core values that guide your counseling practice? How might those values be threatened or enhanced by AI tools?

5. What non-verbal or emotional cues have helped you understand a student's unspoken needs—cues that no data dashboard could detect?

6. How does your school currently balance the use of AI tools and human-led interventions? Is that balance appropriate?

7. In your experience, what are the risks of relying too heavily on digital solutions for student mental health or academic support?

8. How do you personally "hold space" for students? What practices or mindsets allow you to be fully present?

Action Steps

1. Reflect on your recent interactions with students and identify one that demonstrates the importance of human connection. Consider journaling about what made it impactful and how technology either supported or detracted from it.

2. Review any AI tools currently used at your school for mental health, early alerts, or academic advising. Ask: Does this tool support or replace counselor interaction? Bring your reflections to a team meeting or leadership conversation.

3. Share the case study from this chapter (the emotional wellness app) during your next staff meeting to spark discussion on human oversight in AI tool implementation.

4. Create a student feedback loop by asking a few students how they experience support from technology versus in-person counseling. Use their input to advocate for student-centered policies.

5. Conduct a quick check-in with yourself and your team: Are we using tech to create more time for human connection, or is it unintentionally replacing connection?

6. Set a professional boundary or reminder to protect time for one-on-one, relationship-based counseling, even amidst administrative or tech-related demands.

7. Develop a protocol for counselor follow-up when an AI system flags a student concern. Ensure that digital alerts are always paired with human response.

8. Advocate for narrative documentation in addition to data tracking. Encourage counselors to capture students' lived experiences and context beyond what algorithms can record.

Chapter 4

Ethical Imperatives in the Use of AI

As AI becomes more integrated into schools, counselors find themselves at an ethical crossroads. On one side lies the promise of improved efficiency, early interventions, and broader access to support. On the other side lies the risk of surveillance, bias, and harm, especially for the students most in need of protection.

Counseling has always been guided by a robust ethical framework. But how do we apply those principles in a world where student data is interpreted by machines? Where mental health tools operate 24/7 without oversight? Where decisions are made by algorithms trained on incomplete or biased datasets?

The Artificial Intelligence and the Future of Teaching and Learning report (U.S. Department of Education, 2023) offers critical insights that guide ethical considerations around how AI systems should handle student data. Its recommendations align with a larger vision: that AI must serve students, not the other way around.[9]

As AI becomes a more powerful force within education, ethical data collection cannot be an afterthought. It must be the foundation upon which AI systems are built.

Drawing from national guidance and grounded in educational values, ethical AI data practices demand:

- Protecting student privacy and dignity,
- Eliminating bias and discrimination,
- Ensuring transparency and accountability,
- Centering human judgment and relational trust,
- And involving students and educators directly in shaping AI systems.

The future of AI in education must not prioritize efficiency or predictive power over humanity. Our technologies must affirm what every counselor, educator, and student already knows: trust, care, and agency are not optional; they are the heart of learning. Counselors must rise to meet this moment.

This chapter outlines the core ethical responsibilities of counselors and how they must evolve to address the challenges of artificial intelligence.

Revisiting Core Ethical Principles

Professional counseling ethics are shaped by organizations such as the American School Counselor Association (ASCA) and the American Counseling Association (ACA). Core principles include:

- Confidentiality and privacy
- Informed consent
- Equity and non-discrimination
- Beneficence and nonmaleficence (doing good and avoiding harm)
- Professional competence

These principles must now be expanded and reinterpreted to include AI-related issues such as data transparency, algorithmic accountability, and digital consent.

The Ethics of Data Collection

AI systems rely on data: grades, attendance, behavior records, language use, even mood inputs. But students and often parents are not fully aware of how this data is being collected, stored, or interpreted.

Ethical questions to ask include:

- Who owns the student data?
- Can students opt out of AI systems?
- Is data being used only for educational benefit or also for discipline, funding, or profiling?

In the report conducted by the Center for Democracy & Technology (CDT), "Off Task: EdTech Threats to Student Privacy and Equity in the Age of AI," CDT found that 73% of parents expressed concerns about their child's school's data privacy and security practices, a notable increase from previous years.[10] While this statistic doesn't directly reference AI-powered tools, it underscores a growing apprehension among parents about how student data is managed in the context of emerging educational technologies.

Counselors must advocate for transparent data practices and ensure students and families understand how their information is being used. This includes pressing schools and vendors for plain-language data policies and informed digital consent forms.

The Problem of Bias

AI systems are only as fair as the data used to train them. If a predictive model is trained on historical data that reflects racial or socioeconomic disparities, it may reinforce those same inequities.

Example: An AI model that flags students for "disciplinary risk" may disproportionately target Black and Latino boys based on biased historical records, not current behavior.

A 2024 RAND Corporation study found that school districts serving higher-poverty and more diverse student populations were significantly less likely to provide AI training or guidance, raising concerns about whether current AI tools adequately meet the cultural, linguistic, and socioeconomic needs of all students.[11]

"Unlike human counselors, AI lacks the ability to holistically consider a client's complex personal history, cultural context, and varied symptoms and factors among others," the guidelines state. "Therefore, while AI can be a supportive tool, it should not replace the professional judgment of professional counselors. It is recommended that AI be used as an adjunct to, rather than a replacement for, the expertise provided by professional counselors."[12]

Counselors must recognize that bias in AI is not a glitch—it's a mirror of systemic inequities. Ethical counseling requires questioning the sources of data, demanding validation studies, and resisting reliance on tools that have not been audited for bias.

Student Autonomy and Consent

AI tools often run in the background, automatically analyzing student work, mood check-ins, or online behavior. But meaningful student autonomy demands that students understand what is happening and have a say in the process. This includes being informed about how their data is used, how decisions are made on their behalf, and being empowered to ask questions, offer input, and make choices that reflect their values and needs.

Counselors should ensure:

- Students are informed participants, not passive subjects
- Consent is revisited regularly, especially with minors
- Parents and guardians are part of the conversation where appropriate

If a tool provides mental health advice, students must know the difference between automated suggestions and professional care. Misunderstandings here can be dangerous.

Confidentiality in a Digital Era

Counselors are trained to protect student privacy. But when AI systems generate reports or alerts, that information may be stored in vendor databases, shared with administrators, or made accessible to third parties.

Key concerns include:

- Who has access to sensitive AI-generated insights?
- Can a student's private concern become a permanent digital record?
- Are schools complying with FERPA, HIPAA, or state-specific laws?

The growing use of AI in education highlights an urgent need for clear, enforceable guidelines regarding the storage, access, and protection of AI-generated student records. Counselors must take on the role of digital privacy advocates, collaborating closely with IT departments and institutional leaders to ensure that ethical standards for confidentiality and data protection are consistently upheld within digital systems.

Professional Competence and Tech Literacy

Ethical counseling also means staying competent and embracing lifelong learning. As AI becomes more common, counselors must take active steps to develop tech literacy:

- Understand the basics of how AI tools function
- Ask critical questions before implementing new systems
- Seek professional development and cross-sector collaboration

In 2023, the American Counseling Association (ACA) released Integrating AI and LLMs into Counseling Education: Ethical and Inclusive Approaches, calling for a stronger integration of technology into counselor training. The report emphasized the urgent need for counseling programs to address artificial intelligence, digital ethics, and data literacy, highlighting that many programs are still in the early stages of preparing future counselors for the ethical challenges and opportunities presented by AI-driven environments.[13]

Ethical leadership means not only knowing when to use a tool, but also when not to. It requires discernment and a deep understanding of context. In an age of rapid technological advancement, ethical leaders must critically assess whether a tool aligns with student well-being. Sometimes, the most responsible choice is to pause, question, or opt for a more human-centered approach, especially when the risk of harm outweighs the promise of efficiency.

Building Ethical Safeguards

Before adopting any AI-driven tool, institutions, schools, and districts should be guided by an ethical evaluation process. Counselors can lead or inform this effort by asking:

- What is the tool's intended purpose?
- Was it designed with equity in mind?
- Who was involved in its development? Were counselors or students consulted?
- How is data stored, and who can access it?
- What's the plan if something goes wrong (e.g., a false mental health flag)?

Ethics must be proactive, not reactive. That means embedding human judgment, cultural responsiveness, and relational care into every decision involving AI.

Counselors as Ethical Stewards

Counselors are uniquely positioned to lead conversations about ethics in AI. While technologists build systems for scalability, counselors work at the level of the individual. This human-centered lens is exactly what schools need to ensure technology serves students, rather than sorting, labeling, or overlooking them.

The ethical challenges of AI are real. But so is the opportunity for counselors to guide its implementation with empathy, fairness, and integrity.

In the next chapter, we'll dive deeper into one of the most pressing ethical imperatives of our time: ensuring that AI use in institutions is equitable, particularly for marginalized and underserved students.

Reflection Questions

1. How confident do you feel in your understanding of student data rights and digital consent? What areas do you want to learn more about?

2. Have you ever raised concerns about the ethics of a tool or system being used at your school? What was the outcome?

3. Reflect on a time when a student's privacy or autonomy may have been compromised by technology. How did you respond? How would you respond differently now?

4. Which core counseling ethic—confidentiality, consent, equity, or nonmaleficence—feels most vulnerable in the current digital landscape? Why?

5. In your school or district, who makes decisions about adopting new AI or EdTech tools? Are counselors meaningfully involved in that process?

6. How do you currently explain the use of digital tools to students and families? Are those conversations clear, equitable, and empowering?

7. What does it mean to you to be an "ethical steward" in a time of rapid technological change? How does that show up in your daily practice?

8. What biases might be embedded in the data your school uses to flag at-risk students? Who is most likely to be misrepresented or overlooked?

Action Steps

1. Review the ethical guidelines from ASCA and ACA with a focus on how they apply to AI and data-driven tools. Highlight areas where your school's current practices may fall short.

2. Request a meeting with your school or district's IT or data team to better understand how student data is collected, stored, and used in AI-powered tools. Ask who has access and how long data is retained.

3. Create or update a student and family friendly consent form that explains in plain language how AI tools are being used and what rights students have regarding their data.

4. Initiate a discussion with your school leadership about developing an AI tool review protocol that includes counselor input, cultural bias auditing, and student feedback.

5. Compile a list of AI-powered tools in use at your site and assess each one using the following criteria: purpose, equity design, data use, consent clarity, and counselor involvement.

6. Facilitate a professional development session or ethics roundtable with fellow counselors to explore case studies of AI use in education. Reflect on where ethical boundaries may blur.

7. Identify one vendor or platform used at your school and research its privacy policy, data sharing agreements, and any known bias reports or audits. Share your findings with your team.

8. Commit to one tech-related training or certification this year (e.g., data privacy, AI in education, digital ethics) to increase your professional competence and advocate more effectively.

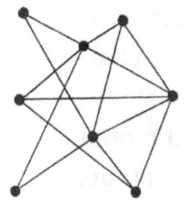

Chapter 5
AI Through a Cultural and Equity Lens

Technology is not neutral. Artificial intelligence is often framed as objective and impartial. But the truth is, technology reflects the values, assumptions, and biases of the people who build it. In education, especially in counseling, this matters profoundly.

Counselors serve students with diverse racial, ethnic, cultural, linguistic, gender, and socioeconomic identities. These students bring unique lived experiences, needs, and worldviews that must be affirmed. When AI tools are introduced in educational settings and institutions without an equity lens, they risk flattening student identity into data points, overlooking critical context, and reproducing systemic harm.

Ethical counseling in the age of AI requires a critical lens, one that questions not only how algorithms are used, but whether they align with the values of relational, student-centered practice. Counselors must remain vigilant in ensuring that data-driven tools do not displace professional judgment or unintentionally replicate systemic bias. To uphold equity and student well-being, counselors should be equipped to assess algorithmic tools, challenge opaque or exclusionary systems, and advocate for AI designs that are transparent, accountable, and rooted in justice. This includes calling for routine equity audits and institutional guardrails that prevent harm.

As AI becomes increasingly embedded in educational infrastructure, the inclusion of counselors in AI governance and decision-making processes is essential, not optional. Despite their central role in supporting students, counselors are too often excluded from conversations about the deployment of AI tools in schools, a gap that risks widening educational disparities rather than closing them. Artificial intelligence continues to reshape education, and algorithmic systems are being increasingly integrated into everything from admissions and grading to early warning systems and personalized learning platforms. However, this integration carries a critical risk: algorithmic bias, which threatens to amplify rather than eliminate existing educational disparities, particularly for students from marginalized communities.

A 2025 study by Boateng and Boateng underscores how algorithmic bias can manifest at multiple stages of AI implementation, from data collection and design to decision-making and deployment. These biases are not random; they disproportionately affect students by race, gender, disability, nationality, and socioeconomic status.[14]

For instance:

- Admissions algorithms that rely on historical data often reinforce racial inequities, especially when race-based considerations are removed under legal constraints.

- Automated grading systems, such as Automated Essay Scoring (AES), have been shown to reflect rater biases or penalize students from underrepresented backgrounds.

- Predictive analytics tools frequently misidentify students of color as "at risk," resulting in fewer supports or even punitive interventions rather than equitable guidance.

These biased systems risk creating a self-reinforcing cycle of inequality. If left unchecked, they can influence long-term educational outcomes, college and career access, and students' sense of agency, subtly enacting a new form of digital redlining in education. This is particularly dangerous in counseling contexts, where ethical practice hinges on human-centered support, trust, and personalized advocacy.

When Systems Reflect Inequities

Bias doesn't begin with AI. It exists in the education system itself—in discipline practices, academic tracking, access to college readiness resources, and even instructor and teacher expectations. When AI is trained on this history, it inherits the same inequities.

Examples of inequitable AI outcomes in education:

- Predictive models can disproportionately flag students of color as "at-risk."
- English learners may be excluded from college-matching algorithms.
- AI-powered mental health tools can ignore cultural norms around emotional expression.
- Career planning tools may steer girls or students from low-income backgrounds toward limited options.

Unless we intervene, AI will amplify the inequities it is fed.

Cultural Responsiveness and AI

Culturally responsive counseling centers the student's unique identity in every conversation. It recognizes that each student's reality is shaped by culture, community, values, and history. But AI systems lack this sensitivity. For example, they don't understand:

- The impact of immigration status on stress
- How cultural norms affect help-seeking behaviors

- The trauma of discrimination or generational poverty
- The language nuances that influence how students express emotion

This is where counselors must act as interpreters and advocates. When using or evaluating AI tools, counselors must ask: "Whose culture is this tool designed for? And who does it overlook?"

Equity-Driven Evaluation of AI Tools

Before adopting a new technology, counselors should help lead an equity audit by asking:

- Was the tool tested on students from diverse racial, cultural, and linguistic backgrounds?
- Are there mechanisms to flag and correct biased outputs?
- Does the tool allow students to define themselves beyond pre-programmed categories?
- Can the system be personalized or adapted to reflect local community needs?

The U.S. Department of Education's 2023 report titled "Artificial Intelligence and the Future of Teaching and Learning" emphasizes that many AI tools are being implemented in educational settings without thorough evaluation of their potential impacts on equity. This lack of comprehensive assessment risks exacerbating existing disparities among students, particularly those from historically marginalized communities. The report calls for intentional, equity-centered practices to guide the development, testing, and adoption of AI technologies in education.[15]

When vendors cannot answer these questions or resist them, that is a red flag.

Co-Designing with Communities

Equity isn't something done to communities, it must be done with them. This includes:

- Involving students and families in the decision to adopt AI tools
- Hosting listening sessions to understand their concerns and hopes
- Prioritizing transparency, language accessibility, and informed choice
- Gathering feedback from underrepresented groups early and often

According to a 2023 study, 73% of parents expressed concern about the privacy and security of their children's educational data. Additionally, 89% of parents stated that they would want to be notified if their child's school were considering the use of artificial intelligence or automated algorithms to make decisions regarding student learning or educational opportunities.[16]

By engaging communities, counselors can ensure that AI reflects the needs and values of those it is meant to serve.

Protecting Vulnerable Populations

AI can create unintended harm when used without equity safeguards, especially for:

- Students with disabilities: May be misclassified or excluded from services
- LGBTQ+ students: May not feel safe disclosing identity to automated systems
- Immigrant or undocumented students: Data collection may trigger fears around privacy or deportation
- Trauma-affected youth: Algorithms may misread behaviors or emotional signals

Counselors must advocate for trauma-informed, inclusive, and flexible systems that protect student dignity and avoid harm.

A Framework for Equitable AI Use in Counseling

Counselors can adopt the following framework when evaluating or implementing AI:

- **Awareness**
  Understand how systemic inequities show up in data, algorithms, and design.

- **Reflection**
  Ask whose voices are missing and whose experiences are centered.

- **Partnership**
  Engage students and families in tool selection and evaluation.

- **Advocacy**
  Challenge vendors and school policymakers to prioritize equity.

- **Action**
  Use data ethically, always paired with compassion and cultural insight.

Equity is a Practice, Not a Checkbox

Equitable use of AI is not a destination—it is a continuous practice that must be revisited often, especially as student demographics and technologies evolve. It requires counselors to be culturally humble, critically curious, and courageously honest about when tools are doing more harm than good.

In the next chapter, we'll look at how to strike a balance between technology and the human role in counseling by exploring practical models for human-AI collaboration that center student care.

Reflection Questions

1. How do the AI tools currently used at your school reflect or fail to reflect the cultural and linguistic diversity of your student population?

2. Have you ever seen a student mischaracterized or misserved by a data system or algorithm? What happened, and how did you respond?

3. What cultural assumptions might be embedded in the educational technology you're asked to use? Who benefits from these assumptions and who might be harmed?

4. How do you ensure that your own counseling practice is culturally responsive? How does that commitment translate to your use of digital tools?

5. Do students and families in your community know how technology is used in schools to track or support them? Are they invited to give input or raise concerns?

6. How do your students express distress, resilience, or identity in culturally specific ways? Could an AI system misinterpret those signals?

7. What groups in your school community may be especially vulnerable to harm or exclusion from AI-driven tools? How are their needs currently being addressed?

8. What does equity-driven technology use look like in action? What would need to change at your school or district to move in that direction?

Action Steps

1. Conduct an informal equity audit of one AI tool currently used at your school. Ask: Who was it designed for? Whose experiences are centered and whose are missing?

2. Start a dialogue with students or families about their experiences with school technology. Consider hosting a listening circle or student-led feedback session.

3. Develop a checklist of equity-focused questions to ask when evaluating any new AI platform (e.g., cultural relevance, language flexibility, bias auditing).

4. Review disaggregated student data from your school's AI-based early warning or academic tracking systems. Look for patterns of over-flagging or underrepresentation.

5. Partner with an equity-focused colleague or team (such as an EL coordinator, DEI officer, or parent liaison) to co-review AI adoption plans or vendor pitches.

6. Create an "Equity and Tech Watchlist": a living document where you and colleagues track concerns, exclusions, or red flags in current platforms.

7. Advocate for co-design practices: Ensure students, families, and frontline counselors are involved early in tech selection and evaluation, especially from underrepresented communities.

8. Share relevant research findings (such as from Pew, RAND, or the AI Now Institute) with leadership to inform policy and push for equity-centered decision-making.

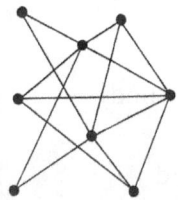

Chapter 6
Collaboration, Not Replacement

The introduction of AI into education has sparked both excitement and concern. On one hand, it promises efficiency and broader reach. On the other, it raises fears about counselors being sidelined or replaced by machines. But here's the truth: AI cannot replace the heart of counseling; it can only support it.

This chapter explores how counselors can thoughtfully integrate AI as a collaborative partner in their work—not as a substitute, but as a tool that enhances their ability to serve students. With intention and clear boundaries, AI can lighten administrative burdens, improve responsiveness, and strengthen student engagement, while preserving the deep human connection that counseling requires.

The Myth of Replacement

AI excels at pattern recognition, automation, and surface-level interaction. But it lacks the intuition, emotional intelligence, and contextual awareness that define counseling.

AI cannot:

- Detect subtle shifts in a student's tone or body language
- Understand the cultural meaning behind certain behaviors
- Navigate complex ethical decisions
- Offer unconditional positive regard or therapeutic presence

Replacing counselors with AI is an unethical goal. But collaborating with AI, when done right, can be a powerful way to expand capacity and deepen impact.

A widely cited 2020 article by McKinsey & Company, "How Artificial Intelligence Will Impact K–12 Teachers," highlights the significant potential for technology to reshape educators' roles. The report estimates that 20 to 40 percent of current teacher tasks could be automated using existing technologies, freeing up approximately 13 hours per week. This time could then be redirected toward activities that directly improve student outcomes and enhance teacher satisfaction. Furthermore, McKinsey suggests that technology could allow teachers to reallocate 20 to 30 percent of their time toward more impactful, student-centered learning activities.[17]

"There appears to be a consensus among teachers and students that generative AI cannot replace the human qualities of teachers that are essential for facilitating students' generic competency development and personal growth....

Ultimately, creating a synergy between humans and technology is key to success in an AI-dominated world."[18]

Framework of Human–AI Collaboration

Let's explore a few frameworks where human expertise and AI tools can complement each other in K16 counseling settings:

Augmented decision-making
- How it works: AI analyzes trends (e.g., attendance dips, academic struggles) and flags students who may need support.
- Counselor's role: Reviews flagged cases, adds context, and follows up with personal outreach.
- Benefit: Allows proactive intervention while retaining human judgment.

Workflow automation
- How it works: AI systems automate repetitive tasks like appointment reminders, scheduling, and progress tracking.
- Counselor's role: Focuses energy on direct student interaction.
- Benefit: Reduces administrative overload, creating more time for high-impact counseling.

Supplemental communication

- How it works: AI chatbots provide basic answers or check-ins after hours.
- Counselor's role: Reviews logs, follows up as needed, and monitors tone and patterns over time.
- Benefit: Provides students with 24/7 access to low-level support while keeping the counselor informed.

Personalized student planning

How it works: AI-powered tools suggest academic or career pathways based on student profiles.

- Counselor's role: Interprets the recommendations through a cultural and developmental lens.
- Benefit: Helps counselors tailor plans more efficiently while preserving student agency.

Guardrails for Ethical Collaboration

To ensure AI stays in a supportive role rather than a dominant one, counselors must set clear boundaries:

- **Human first, tech second**
  All decisions should center the counselor–student relationship.

- **AI as a mirror, not a map**
  Use AI insights as starting points for inquiry, not conclusions.

- **Transparent communication**
  Let students and families know when AI is being used and how.

- **Data review and context**
  Always pair AI outputs with professional interpretation.

A Shared Vision for Implementation

Successful human–AI collaboration depends on clear alignment between counselors, administrators, IT teams, and tool developers. Here's how to foster that alignment:

- **Define the purpose**
  What problem are we trying to solve? Can AI help?

- **Build with counselors, not just for them**
  Include counseling professionals in selection, piloting, and evaluation processes.

- **Ensure ongoing support**
  Provide training and feedback channels for continued improvement.

- **Commit to equity**
  Ensure AI serves all students, especially those historically underserved.

Case in Practice: Augmented Counseling, Not Automated Care

At one middle school, the counseling team worked with an EdTech company to pilot a self-reflection app for students. The app used AI to identify emotional trends and flag students who appeared distressed. Instead of letting the system act autonomously, counselors set up a workflow where they reviewed flagged entries each morning.

The result? Counselors were able to reach out faster to students in need, and students appreciated knowing a real person was reading and responding. It was AI with empathy, not in place of it.

A Partnership, Not a Power Shift

The goal of AI in counseling should be to amplify humanity, not replace it. When AI and counselors work in partnership, each doing what they do best, students benefit from timely, personalized, and compassionate support.

In the next chapter, we'll dive into how counselors can build the competencies they need to lead in this new era, including training pathways and strategies for becoming AI-literate without losing their human-centered roots.

Reflection Questions

1. Have you felt pressure (explicit or implied) to rely on technology for tasks that used to be done through personal interaction? How did that affect your role?

2. In what areas of your counseling work do you feel overwhelmed by administrative tasks? Could AI help free up time for deeper student engagement?

3. Have you been involved in selecting or implementing any AI tools at your school? If not, how might your input have improved the process?

4. What do you think about the idea of AI offering "recommendations" for student planning? How could you ensure those recommendations respect student identity and context?

5. What boundaries would you personally want to set when it comes to AI interacting with students (e.g., messaging apps, chatbots, or mood check-ins)?

6. What does successful human–AI collaboration look like to you in a counseling context? What values should guide that partnership?

7. How do you communicate with students and families about the role of AI in your work? Is there a need for more transparency or clarity?

8. What would help you feel more confident in using AI tools ethically and effectively in your role?

Action Steps

1. Identify one current task in your counseling workflow that could be ethically supported by AI (e.g., reminders, follow-ups, check-ins). Research whether a tool already exists that aligns with your values.

2. Map out your AI boundaries and create a list of counseling functions you believe should never be automated and others where support tools may be appropriate. Share this with your team or leadership.

3. Meet with your site or district IT team to discuss how AI is being implemented and whether counselors are part of the planning, piloting, and review process. Request a seat at the table.

4. Audit one AI-driven system at your school and evaluate whether its outputs are being reviewed by a counselor. If not, recommend a system for human follow-up and contextual interpretation.

5. Develop a "Human First" protocol to guide how AI insights are reviewed, responded to, and communicated with students and families. Make sure the counselor's role is clearly defined.

6. Create a presentation or handout for students and parents explaining how AI is used in counseling at your school—what it can do, what it cannot, and how human oversight is maintained.

7. Host a team discussion or workshop using the frameworks in this chapter (augmented decision-making, workflow automation, etc.) to brainstorm where AI could support your existing systems.

8. Advocate for professional development around AI in education with a focus on ethical implementation, collaboration frameworks, and student-centered design.

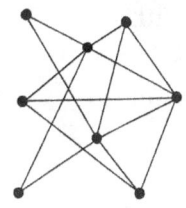

Chapter 7

Developing AI Literacy for Counselors

In an increasingly digital world, counselors are expected to support students navigating both personal challenges and evolving technologies. But while AI tools are being rapidly adopted across institutions and school systems, many counselors have not been trained to understand, question, or shape these tools.

AI literacy doesn't mean becoming a coder or data scientist. It means building the confidence and critical thinking skills to ask the right questions, understand the tools being used, and advocate for ethical and equitable implementation.

While the use of AI in education is rapidly expanding, there is a notable lack of specific data regarding counselors' training and preparedness in this area. Although several national surveys highlight efforts to train teachers on AI technologies, few studies focus explicitly on counselors. This gap in research raises important concerns, as counselors are often critical in guiding students through increasingly technology-driven educational environments. Without targeted training and support, counselors risk being left out of vital conversations about AI implementation, data ethics, and student advocacy.

This lack of preparedness can lead to the unethical adoption of tools that may unintentionally harm or misrepresent students.

This chapter provides a roadmap for developing AI literacy within the counseling profession, focusing on knowledge, skills, and strategies that keep students at the center.

What Is AI Literacy for Counselors?

AI literacy includes:

- Basic understanding of how AI works (e.g., algorithms, machine learning, natural language processing)
- Awareness of limitations and risks, including bias, privacy concerns, and dehumanization
- Critical evaluation of tools and data sources
- Ethical discernment about appropriate uses in counseling
- Advocacy skills to influence tech adoption in student-centered ways

AI-literate counselors don't just react to new technologies; they shape how they're used, protecting student well-being and advancing equity.

The Foundations: What Counselors Should Know

To build AI literacy, counselors should begin with these essential concepts:

- **What is AI?**
  At its core, AI is a system that mimics human intelligence using algorithms and data. It "learns" patterns to make decisions or predictions.

- **What is Machine Learning (ML)?**
  ML is a type of AI where systems improve performance over time based on data inputs. In schools, ML might be used to predict dropout risk or flag behavior concerns.

- **What is Natural Language Processing (NLP)?**
  NLP helps machines understand and respond to human language—powering tools like chatbots, sentiment analysis, or text-based wellness check-ins.

- **What Are Algorithms and Data Sets?**
  Algorithms are sets of rules or calculations. They learn from data—so if the data is biased, the output will be too.

Professional Development Opportunities

Counselors can build AI literacy through a variety of pathways:

- **Workshops and webinars**
  Attend training hosted by EdTech organizations, universities, or professional associations focused on AI in education.

- **Collaborate with EdTech teams**
  Partner with IT staff and technology specialists in your district. Ask to review tools being implemented or attend vendor demos.

- **Self-paced learning**
  Explore resources like:
  —Google's Machine Learning Crash Course
  —Common Sense Education: AI Literacy Toolkit
  —ASCA and ISTE webinars on AI and student data

- **Higher education coursework**
  Some counseling and education graduate programs now offer modules on technology in

student services, an ideal space to deepen your knowledge.

Key Questions AI-Literate Counselors Ask

Before using or supporting any AI tool, counselors should consider:

- What problem is this tool trying to solve?
- How was it trained—and on whose data?
- Who was involved in its development?
- What are its limitations?
- How will students and families be informed or involved?
- What happens if the tool gets it wrong?

These questions shift counselors from passive users to ethical evaluators and co-creators of AI-enabled systems.

Building a Culture of Inquiry

AI literacy is not just about personal learning, it's about building collective capacity. Counselors can:

- Form a learning circle with colleagues to explore new technologies together

- Host equity and ethics conversations before any new tech is adopted
- Invite student input when evaluating tools
- Push for inclusive design in district-level decisions

When counselors model curiosity, collaboration, and critical thinking, they help create a school culture where innovation and integrity go hand in hand.

Partnering with Others

Counselors don't need to do this alone. Building AI literacy means forging partnerships with:

- District IT teams and digital learning leaders
- Professional organizations like ASCA, ACSA, or ISTE
- University researchers working on ethical EdTech
- Students and families, who bring firsthand experience and insight

These collaborations can lead to more thoughtful implementation, shared learning, and stronger advocacy for human-centered design.

Confidence, Not Perfection

You don't need to know everything about AI to begin making an impact. AI literacy is a journey, not a destination. What matters most is the willingness to learn, ask questions, and stay rooted in counseling values.

The next chapter will explore real-world applications, including case studies of schools using AI tools effectively and ethically, as well as a step-by-step guide for counselors leading responsible implementation.

Reflection Questions

1. How confident do you feel in your understanding of how AI tools used in your school function? What are your biggest knowledge gaps?

2. Have you ever used an AI-powered tool in your counseling practice? If so, how did you assess its effectiveness and ethical implications?

3. What concerns do you have about AI tools being implemented without counselor input? Have you witnessed this happen at your site?

4. What does AI literacy mean to you personally and how could it enhance your role as a student advocate?

5. When evaluating new technology, do you feel empowered to ask critical questions about ethics, bias, and data usage? Why or why not?

6. How does your school or district currently train staff on AI and emerging technologies? What would you improve about this process?

7. What partnerships, inside or outside your school, could help you grow your AI knowledge and confidence?

8. In what ways can you model AI literacy for your students, colleagues, or administrators?

Action Steps

1. Complete a short AI literacy module such as Google's Machine Learning Crash Course or a relevant Common Sense Education toolkit to build foundational understanding.

2. Create a "cheat sheet" of key AI terms (e.g., algorithm, NLP, bias, data set) and post it in your office or digital space as a quick reference for yourself and your colleagues.

3. Organize a learning circle with fellow counselors to read articles or watch short webinars together and reflect on implications for student support.

4. Attend an AI-related PD session or webinar offered by organizations like ASCA, ISTE, or your local county office of education. Share your takeaways with your counseling team.

5. Request to join or observe your district's EdTech evaluation committee or vendor demo meetings, advocating for counselor representation.

6. Practice asking the six key AI questions (from the chapter) when reviewing any tech tool used for student support. Share these questions with administrators or IT staff to prompt deeper evaluation.

7. Invite a digital learning specialist to a counseling department meeting to provide an overview of AI tools currently being piloted and offer time for open discussion.

8. Design a short workshop or resource guide for colleagues titled "What Every Counselor Should Know About AI" to build collective capacity at your site.

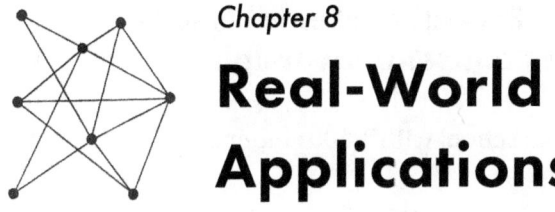

Chapter 8

Real-World Applications

By now, we've explored the principles, ethics, and frameworks for integrating AI into counseling in ways that preserve human connection and advance equity. But what does this look like in practice?

This chapter highlights real-world examples of AI use in counseling settings, both promising and problematic. It also offers a step-by-step implementation guide for counselors and educational leaders to introduce or evaluate AI tools in a student-centered, ethical way.

Despite the rapid expansion of AI in education, there remains a critical and troubling gap: counselors are largely being left out of AI training initiatives. While efforts have been made to equip teachers, little attention has been paid to preparing counselors for the ethical, academic, and emotional challenges that AI introduces into student support systems. Without immediate action to provide targeted training, counselors risk losing their voice in key conversations about data privacy, equity, and the responsible use of technology in schools—at the very moment when students need their guidance the most.

Case Study 1: Proactive Mental Health Support with Counselor Oversight

Setting: Urban high school with 1,500 students

Tool Used: AI-powered journaling app with mood detection

Challenge: High student-to-counselor ratio; students reluctant to request emotional support

Solution:
- Students opt in to use a journaling app that checks in daily with open-ended prompts.
- The app uses Natural Language Processing (NLP) to identify students showing signs of distress (e.g., hopelessness, anxiety).
- A dashboard highlights concerning entries—but only to the counseling team, not administrators.
- Counselors use insights to guide personalized outreach and invite students to check-ins.

Outcome:
- Increased early interventions, especially for students who hadn't previously accessed services.
- The school formed a student data committee to co-create privacy safeguards.

Lesson: AI insights must be paired with human follow-up and student agency.

Case Study 2: Career Pathways Platform with Built-In Bias

Setting: Rural middle school using a national AI career-matching tool

Tool Used: Predictive pathway generator based on student assessments and behavior logs

Challenge: Discovered that the tool frequently steered female students toward caregiving careers and underrepresented minority students toward low-wage job tracks

Response:

- Counselors analyzed outcomes and flagged patterns to the vendor.
- School paused use and conducted a local review with parents and community partners.
- A new model was implemented allowing students to manually adjust suggestions and explore careers not generated by the algorithm.

Lesson: Without continuous equity audits, AI can reinforce stereotypes.

Case Study 3: Automating Administrative Work for Counselor Efficiency

Setting: Suburban K–8 school with one counselor serving 800 students

Tool Used: AI scheduling assistant and automated case notes

Goal: Reduce administrative workload to allow more face-to-face counseling

Approach:

- The counselor used a secure, FERPA-compliant system to schedule meetings, generate summaries, and flag follow-up reminders.
- Confidential information was restricted from automation.
- Time saved was redirected into implementing new small group sessions and classroom guidance lessons.

Outcome: The counselor doubled time spent in direct student support while maintaining documentation compliance.

Lesson: When used strategically, AI can expand, not replace, relational work.

Case Study 4: Supporting Equitable Transfer Pathways Through AI-Enhanced Advising, Community College Transfer Center, and University Advising Office

AI Tool: EduNav (by Civitas Learning)

Context: Celia is a first-generation Latina student at a California community college majoring in psychology with plans to transfer to a CSU. She's balancing part-time work, caregiving responsibilities, and a full course load. Celia is eligible for a transfer guarantee program but is unaware of the process and deadlines.

Challenge: Despite consistent academic performance, Celia has not received timely guidance on her transfer options. The college's counseling department is overextended, and her questions about articulation agreements and GPA calculations remain unanswered.

Intervention: The college implements EduNav, an AI-powered degree planning and advising platform that integrates with the college's Student Information System (SIS). EduNav analyzes student data in real time and offers predictive course planning, milestone alerts, and personalized guidance.

Counselor Action: EduNav identifies Celia as at-risk for missing key transfer milestones. A counselor is notified and reaches out to schedule a meeting. During their session, the counselor uses EduNav to:

- Review Celia's completed IGETC and CSU GE courses and remaining requirements
- Generate a transfer timeline based on her declared major and preferred CSU campuses
- Explore articulation agreements and identify partner universities with robust student support and scholarship opportunities

Outcome: Celia completes her CSU application on time and is accepted to CSU Fullerton with a scholarship. She continues using EduNav at the university level, where it supports her with course sequencing and access to relevant academic resources.

Counselor Reflection: AI-enabled tools like EduNav enhance proactive advising by identifying students who need timely intervention. It didn't replace the counselor's role—it empowered more personalized, strategic support and reduced administrative overhead, allowing counselors to focus on relationship-centered guidance.

Case Study 5: Supporting Veteran Student Re-Entry with AI-Driven Academic Planning, Veteran Resource Center, Community College Counseling Department

AI Tool: Navigate360 (by EAB)

Context: James is a 34-year-old Marine Corps veteran who recently enrolled at a local community college to pursue a degree in business. After years of military service and sporadic employment, James is returning to school through the GI Bill. While motivated, he struggles to navigate course requirements, military credit transfers, and the complexities of his benefits.

Challenge: James is unsure how his military experience translates into college credits. He also finds it difficult to connect with traditional-aged peers and often misses important administrative deadlines due to juggling school, part-time work, and family responsibilities.

Intervention: The college's Veteran Resource Center collaborates with the counseling department to leverage Navigate360, an AI-enhanced platform that supports onboarding, case management, and guided pathways for nontraditional students.

Counselor Action: Navigate360 flags James as a new veteran enrollee with no declared pathway and pending transcript evaluations. A counselor reaches out to schedule a session. During the meeting, the counselor uses the platform to:

- Review James's military transcripts and match applicable credit to business pathway requirements
- Create a personalized academic plan that aligns with his long-term goals and financial aid timeline
- Automate reminders for key tasks, such as registering for classes, submitting documentation, and attending benefits workshops

Outcome: With tailored guidance and automated nudges, James completes his first semester successfully and builds confidence in his academic identity. He connects with a peer mentor through the platform and continues progressing toward his degree with fewer administrative hurdles.

Counselor Reflection: AI-supported tools like Navigate360 enhance equity for adult learners and veterans by providing structure, timely information, and individualized planning. These features reduce friction in re-entry and allow counselors to build trust and continuity in the advising relationship.

A Counselor's Guide to Responsible AI Implementation

Below is a simple step-by-step framework counselors can follow when introducing or reviewing AI tools:

Step 1: Identify the need

- What student need or counseling gap are you addressing?
- Is AI truly the right solution?

Step 2: Form an evaluation team

- Include counselors, teachers, students, parents, and IT staff.
- Ensure diversity of voices, especially from underrepresented communities.

Step 3: Assess the tool

- Purpose and intended outcomes
- Data sources and ownership
- Equity safeguards
- Cultural responsiveness
- Student privacy and opt-in design

Step 4: Pilot with transparency

- Start small (e.g., one grade level or department)

- Collect both quantitative and qualitative feedback
- Hold student voice sessions to evaluate experience and trust

Step 5: Monitor and adjust

- Don't "set and forget" AI systems, review outcomes regularly
- Track disparities in use or effectiveness
- Make changes based on feedback, not assumptions

Step 6: Share and advocate

- Document successes and challenges
- Advocate for scalable ethical practices across the district
- Collaborate with other schools or regions to build a learning network

From Experiment to Impact

Real-world success with AI in counseling is not about perfect technology. It's about intentional design, ethical leadership, and community trust. Counselors who step into this space equipped with empathy, knowledge, and vision can ensure that technology serves the student, not the system.

In the next chapter, we'll center the voices that matter most: the students. What do they want from AI? What concerns do they have? And how can we co-create support systems that reflect their needs and values?

Reflection Questions

1. Have you ever been asked to use an AI or digital counseling tool without being part of the decision-making process? What was that experience like?

2. Which of the case studies in this chapter resonated most with your school's needs or realities? Why?

3. What systems or processes in your current counseling role could benefit from AI-supported efficiency without compromising your connection with students?

4. How is student voice currently included (or not) in your school's technology adoption or review processes? What would more inclusive involvement look like?

5. Have you ever reviewed an AI tool for equity or bias? What questions did you ask or should you have asked?

6. How comfortable do you feel facilitating conversations with vendors, administrators, or IT teams about student privacy, data ethics, or opt-in design?

7. What steps does your school currently take to monitor the effectiveness and fairness of tools already in use? Could those processes be improved?

8. If you had to advocate for or against an AI tool being introduced at your school today, what evidence or values would guide your stance?

Action Steps

1. Choose one of the six steps in the section "Counselor's Guide to Responsible AI Implementation" and begin working through it with your team. For example, start by identifying a specific student need or gap that might be supported by tech.

2. Conduct a mini-audit of any AI tool currently used in your counseling department. Use the checklist in Step 3, "Assess the tool" (tool purpose, data use, equity safeguards, etc.), as your evaluation rubric.

3. Form or join a cross-functional team (counselors, students, IT staff, teachers, and families) to review or co-design technology initiatives at your site or district.

4. Host a reflection session with students on their experiences with digital tools, especially any that use AI. Use this as an opportunity to gather insights on trust, clarity, ease of use, effectiveness, and consent.

5. Pilot a new process such as AI-assisted scheduling or journaling with a small group. Document outcomes and review feedback regularly.

6. Design a tool review template that your school can use to evaluate future EdTech platforms. Include prompts on cultural responsiveness, bias audits, and data ownership.

7. Share your experiences and outcomes from any AI integration with your district or counselor network. Help others learn from your implementation journey by sharing both the wins and the challenges.

8. Develop a professional learning community (PLC) with counselors across schools or regions to explore best practices in ethical AI use. Schedule quarterly meetups to share tools and support each other's growth.

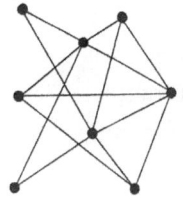

Chapter 9

Student Voice and Well-Being in the Age of AI

When we talk about AI in education, we often hear from administrators, developers, and policymakers. But the people most affected, the students themselves, are rarely part of the decision-making process. This is a critical oversight.

Students are not passive consumers of technology. They are active participants in their education, their mental health journeys, and their futures. As counselors, we must center student voices in every conversation about how AI is used to support or monitor them.

In 2023, the Center for Democracy & Technology published a report titled "Off Task: EdTech Threats to Student Privacy and Equity in the Age of AI." Drawing on surveys of students, parents, and teachers, the report highlights widespread concerns about the use of educational technology and the protection of student data.

A significant majority of students expressed a desire for greater involvement in decisions about how AI systems are used in their schools. Yet despite this strong interest, only a small percentage of students reported being consulted during the implementation of new technologies. This disconnect underscores the urgent need to meaningfully engage students, the individuals most directly impacted, as partners in shaping the future of AI in education.[19]

What Students Are Saying

In student interviews, focus groups, and surveys across the country, key themes have emerged about how young people feel about AI in their schools:

1. **"We want help, not surveillance."**
   Students are wary of AI systems that make them feel like they're being watched. They fear their private thoughts could be misinterpreted, judged, or even shared without consent.

   "Sometimes I feel like I have to censor myself when I use those apps. I don't know who's reading it."

2. **"We want options and transparency."**
   Students want to know what AI tools are being used, how their data is processed, and if they can opt out.

   "At least let us know how it works. We should have a choice."

3. **"We want real people involved."**
   Many students prefer digital tools for convenience but still crave personal interaction.

   "I like the AI app for quick check-ins, but it can't replace someone who actually knows me."

4. **"We want to be involved."**
   Young people want to co-create the systems meant to support them. They ask to be consulted, not just informed.

   "If it's about us, then why weren't we in the room when they made the decision?"

The Emotional Impact of AI

The well-being of students is more than emotional regulation; it's a reflection of how safe and empowered they feel in their learning environment. AI tools can affect this in subtle but significant ways:

Positive Impacts:
- Instant access to self-help tools
- Personalized learning and college/career suggestions
- Reducing stigma through private, digital support

Negative Impacts:
- Feeling over-monitored or mislabeled
- Loss of control over one's own narrative
- Lack of trust in school systems that use AI without explanation

According to the 2024 Global AI Student Survey conducted by the Digital Education Council, concerns about the fairness and impact of AI in education are widespread among students. Sixty percent of students reported being worried about the fairness of AI-driven evaluations, expressing particular unease with how AI is used in teaching, assessment, and grading processes. Additionally, more than half of the students surveyed believe that an over-reliance on AI by educators could negatively affect their academic performance and diminish the overall value of their education.[20]

These findings suggest that while students recognize the potential benefits of AI, they are deeply cautious about its unchecked application. For educational institutions seeking to adopt AI, it will be essential to strike a thoughtful balance—integrating AI tools to enhance learning while maintaining human judgment and relational teaching at the core of the educational experience.

The key is trust, and trust begins with transparency, respect, and inclusion.

Supporting Digital Well-Being

Counselors play a vital role in guiding students not only through life challenges, but through healthy engagement with technology. This includes:

- Teaching digital literacy, how to interpret and question AI recommendations
- Offering space to discuss feelings about data privacy and control
- Helping students navigate algorithmic bias in tools that shape their futures
- Reinforcing that AI does not define their worth

When students understand how AI works—and that they are far more than what it predicts—they feel safer, more empowered, and better supported.

Strategies to Elevate Student Voice

Counselors can take the lead in creating structures where students help shape how AI is used.

- **Host listening circles**
  Create facilitated sessions where students can share their experiences with school technology. Use open-ended questions like:
  —"How does using [tool name] make you feel?"
  —"What would make it safer or more helpful for you?"

- **Establish student tech councils**
  Include students in tool evaluation committees or tech advisory groups. Their feedback can influence selection, implementation, and improvement.

- **Co-create guidelines**
  Work with students to write agreements for how AI tools should (and should not) be used, particularly when it comes to data, privacy, and autonomy.

- **Reflect and respond**
  Always circle back and keep students informed. Show students how their input is being used. If a decision can't be changed, explain why.

Protecting Student Agency

At the heart of student well-being in the age of AI is agency, which is the ability for students to make informed choices about their own lives. This includes:

- Knowing when an AI tool is optional
- Understanding how to correct inaccurate profiles or flags
- Having access to real people when needed
- Being treated with respect, not as data points

Agency is not an optional feature of AI; it's something counselors must protect and nurture.

A Relationship Reimagined

The relationship between students and technology is evolving. But no matter how intelligent our tools become, students will always need people who see them, believe in them, and walk with them. Counselors are those people.

In the final chapter, we'll look toward the future and explore how counselors can lead the movement for ethical, equitable, and human-centered innovation in education.

Reflection Questions

1. How often are students at your school invited to share their perspectives about the technology used to support or monitor them? How could this be improved?

2. Have you noticed students becoming more guarded, anxious, or disengaged when interacting with digital tools? If so, what might be contributing to that?

3. What does "digital well-being" mean in your counseling practice? How do you address it with students?

4. What AI tools in use at your school could affect student self-expression, identity, or sense of trust? How are those tools framed or explained to students?

5. How do you currently support students who are skeptical or fearful of being monitored or mislabeled by AI systems?

6. In your experience, what role do students want to play in shaping how technology is used in their education?

7. What structures exist (or could be created) to elevate student voices in decision-making about EdTech and AI at your school?

8. How do you ensure students know they are more than their data? How do you reinforce that their stories, emotions, and growth cannot be reduced to algorithms?

Action Steps

1. Host a student listening session or "AI & Me" circle. Ask open-ended questions about how students feel using apps, platforms, or check-in tools. Capture their feedback and themes anonymously.

2. Establish or join a Student Tech Council that includes youth in selecting or reviewing AI-powered tools, especially those tied to mental health, behavior, or academic planning.

3. Create a simple visual guide or explainer to share with students and families that breaks down what AI tools your school uses, how data is used, and who sees what. Use plain, accessible language.

4. Develop a "Student Bill of Digital Rights" with a small group of students to outline expectations for consent, transparency, opt-out options, and respectful data use.

5. Pilot a co-created feedback loop: After introducing or using a new tool, invite students to reflect (via survey, discussion, or journal) on how it impacted their trust and sense of being supported.

6. Hold classroom or small group discussions on algorithmic bias and digital identity. Use real-world examples to help students understand how tech shapes perceptions and how to advocate for themselves.

7. Partner with a student club or leadership group to create digital wellness campaigns or peer-led workshops that explore AI, identity, privacy, and self-worth.

8. Reframe counseling check-ins to include tech talk: Ask students how digital tools are impacting their education, how they feel about them, and what they'd change.

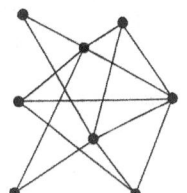

Chapter 10

Looking Forward

AI is no longer just on the horizon. It's here, and it's influencing how institutions deliver services, interpret student needs, and design support systems. But the future is not predetermined. It's shaped by the choices we make today.

This final chapter is a call to action. It challenges counselors to take their seat at the table, not as passive recipients of technology, but as ethical leaders, student advocates, and co-creators of a better, more compassionate educational future.

Three Shifts We Must Lead

To ensure AI strengthens counseling rather than weakening it, we must embrace three critical shifts:

1. **Shift from reaction to leadership**
   Too often, AI tools are introduced without meaningful counselor input. We must move from reacting to tech decisions to proactively shaping them. This means:
   - Asking hard questions early

- Recommending tools that align with student needs
- Serving as ethical gatekeepers and innovation partners

2. **Shift from efficiency to empathy**
   While AI can optimize systems, the purpose of counseling is not speed; it's connection. We must resist the temptation to use technology to replace relationships, and instead harness it to deepen care and understanding.

3. **Shift from one-size-fits-all to equity-driven design**
   Generic tools often fail to meet the needs of diverse students. Counselors can lead the push for culturally responsive, trauma-informed, and inclusive systems that work for every learner.

Becoming Advocates for Ethical AI

Ethical AI requires that people be willing to speak up. Counselors are uniquely positioned to do this work because we understand:

- The complexity of student lives
- The dangers of unchecked data collection
- The importance of context, culture, and care

To become effective advocates, counselors can:

- Engage in policy conversations at the school, district, or state level
- Collaborate with EdTech companies to pilot human-centered tools
- Educate students and families about their digital rights
- Push for funding to support safe, equitable AI practices

As AI becomes increasingly woven into the fabric of education, counselors have a critical role to play in ensuring that technology serves students ethically and equitably. Our deep understanding of student development, lived experiences, and systemic inequities uniquely positions us to lead conversations that might otherwise overlook human needs.

Ethical AI will not emerge on its own; it will require persistent advocacy, courageous conversations, and collaborative leadership. Counselors must be at the forefront, engaging with policymakers, working alongside EdTech developers, empowering students and families to understand their digital rights, and demanding resources to protect student data and well-being.

The future of education will be shaped by those who are willing to advocate for it. As trusted voices within our schools and communities, counselors must step forward to ensure that AI enhances, rather than harms, the human-centered foundations of learning.

Building a Movement, Not Just a Moment

This work will not happen in isolation. It requires collective effort and courageous leadership across the field. Consider forming or joining:

- Regional or national networks focused on AI and education equity
- Cross-sector partnerships with researchers, nonprofits, or tech companies
- Mentorship groups to support other counselors in building AI literacy
- Student advisory councils to keep youth voice front and center

The most powerful innovations are not created in labs—they are co-designed with the communities they serve.

Imagining the Future of Counseling

What might the future look like when counselors lead with intention, empathy, and boldness?

- Students feel seen, supported, and protected rather than surveilled or sorted
- AI tools are trusted because they are transparent, inclusive, and optional

- Families understand how their data is used and how they can engage
- Technology augments care, while counselors remain central to student success
- Schools become models of ethical, equitable innovation

This is not a dream. It is a real and achievable possibility if we choose to pursue it together.

A Final Word

You don't have to be a tech expert to lead in this space. You only need a commitment to students, a willingness to learn, and the courage to speak when it matters most.

The future of counseling in the age of AI is not about choosing between humans and machines. It's about preserving what makes us human while embracing the tools that help us care more deeply, act more quickly, and reach more students with dignity and grace.

Let's make sure we shape this future, not for the sake of innovation alone, but for the lives we are entrusted to guide.

Reflection Questions

1. What mindset shift do you personally need to make to move from reacting to technology to leading its ethical integration?

2. How do you currently ensure that efficiency doesn't come at the cost of empathy in your daily counseling practice?

3. What barriers have prevented counselors from being part of EdTech or AI-related decisions in your school or district? How might those be addressed?

4. How would you define ethical AI in the context of counseling? What values guide your vision of ethical innovation?

5. Who are your allies in this work inside and outside your institution? How might you strengthen those partnerships?

6. What does a student-centered, equity-driven AI tool look like in practice? Can you think of a time when a tool missed the mark? What would you change?

7. How will you continue growing your AI literacy while also supporting your peers and colleagues in this journey?

8. What role do you want to play in shaping the future of counseling? What step can you take today to begin that leadership path?

Action Steps

1. Draft a personal or team "Ethical AI Pledge" that outlines your values and commitments when engaging with AI tools in education. Use it to guide decisions and conversations.

2. Start a policy conversation by presenting to your site or district leadership on the counselor's role in AI implementation. Share highlights from this book and offer clear recommendations for inclusion.

3. Identify one place to lead: Join a district AI task force, apply to a regional network, or propose an equity review process for new tech tools. Bring the counseling lens to the table.

4. Create or co-host a roundtable with tech staff, administrators, and fellow counselors to evaluate whether current AI tools align with student wellness, agency, and equity.

5. Mentor a colleague by sharing what you've learned throughout your AI journey. Offer to facilitate a professional learning session or co-plan an implementation review.

6. Collaborate with student leaders to envision what a future of ethical, inclusive technology should look like at your school. Use design thinking or storytelling formats to amplify their ideas.

7. Design a professional development roadmap for your site or region: What workshops, tools, or discussions could move your team from AI fear to AI fluency and ethical action?

8. Document your work and lessons learned from any AI-related efforts and share it whether at a conference, in a blog post, or with professional organizations. Contribute to the larger movement.

Appendix 1

Student Voice Engagement Toolkit: Co-Creating Ethical and Inclusive AI Practices in Schools

Students are not just users, they are co-creators of school culture. When we honor their insight and center their experiences, we build systems that reflect trust, dignity, and equity for all.

Let's co-create the future. Together.

Purpose of This Toolkit

This toolkit empowers counselors, educators, and student leaders to collaboratively implement AI tools in education ethically, ensuring that student well-being remains central to every decision.

Section 1: Hosting a Student Listening Session

Objective: Create a safe, student-centered space to hear how technology is impacting students' emotional, academic, and digital lives.

Sample questions:

- How do the AI tools or apps used in school make you feel?
- Have you ever felt watched, judged, or misunderstood by a digital platform?
- What would help you feel more informed or in control of how your data is used?
- In what ways can school tech support your well-being better?
- What should adults know about how students use and experience these tools?

Section 2: Forming a Student Tech Advisory Council

Purpose: Give students an ongoing role in evaluating, selecting, and improving digital tools in their schools.

Steps to get started:

- Recruit students from different grades, identities, and tech experiences.

- Meet monthly or bi-monthly to review new or existing tools.
- Train students on AI basics and ethical principles.
- Empower students to co-present findings or recommendations to leadership.

Suggested roles:

- Chair/Facilitator
- Research and Evaluation Lead
- Community Engagement Lead
- Data Ethics Advocate
- Note Taker / Historian

Sample activities:

- Review a wellness app for accessibility and tone.
- Conduct a student survey about tech comfort and concerns.
- Draft a feedback report for administrators.

Section 3: Co-Creating a Student Bill of Digital Rights

Purpose: Empower students to define what ethical tech use looks like in their school centered on agency, transparency, and well-being.

Activity prompt:

- Break students into small groups.
- Ask: What rules would you write to make tech feel safer, fairer, and more useful?
- Invite each group to draft a "mini-bill."
- Combine into a shared declaration.
- Present to school leadership or school board.

Suggested Format:

Our Rights as Students in the Age of AI:

- We have the right to know when AI is being used to make decisions about us.
- We have the right to consent or say no to certain digital tools.
- We have the right to correct or challenge inaccurate data.
- We have the right to real human support when digital systems fall short.
- We have the right to be part of the conversation.

Section 4: Student Follow-Up Planning and Impact Evaluation

Best practices:

- Share how student input was used (or why changes were not possible).
- Set up anonymous follow-up forms or "digital suggestion boxes."
- Offer opportunities for new students to join each semester.
- Invite students to share at PD sessions or tech reviews.

Sample reflection prompt for students:

- "I feel more confident using technology at school because..."
- "One thing I still want adults to understand about AI is..."

Appendix 2
AI Implementation Planning Template

This template is designed to help educators collaboratively evaluate and implement AI tools and practices that are ethical, equitable, and student-centered.

Section 1: Vision & Purpose (Ch. 1, Ch. 10)

- What is the primary goal of introducing AI in your counseling practice or school?

- How does this initiative align with student well-being and human-centered care?

- Who are the stakeholders involved in shaping this vision?

Section 2: Tool Identification & Needs Assessment (Ch. 2, Ch. 8)

- What specific student or system needs are you trying to address?

- What AI tools or platforms are being considered or currently in use?

- Has a gap analysis been conducted? What are your findings?

Section 3: Human Connection & Counseling Priorities (Ch. 3, Ch. 6)

- How will AI be used to enhance, not replace, human connection?

- What core counseling activities must remain human-led?

- What role will counselors play in reviewing or responding to AI outputs?

Section 4: Ethical & Data Review (Ch. 4)

- Who has access to AI-generated student data?

- How is consent gathered and explained to students/families?

- What are your strategies for ensuring data privacy and transparency?

Section 5: Equity & Cultural Responsiveness (Ch. 5)

- Was the tool tested on diverse student populations?

- What equity concerns have surfaced, and how are they being addressed?

- How will you ensure culturally responsive implementation and review?

Section 6: Collaboration & Stakeholder Engagement (Ch. 6, Ch. 9)

- Which departments or partners are involved in implementation?

- How will student and family voices be integrated?

- What communication channels will keep all stakeholders informed?

Section 7: AI Literacy & Capacity Building (Ch. 7)

- What professional development is needed for staff?

- What self-paced or collaborative learning opportunities are planned?

- Who will lead ongoing capacity building efforts?

Section 8: Pilot Process & Feedback Loops (Ch. 8, Ch. 9)

- What is the scope and timeline of the pilot?

- How will feedback be gathered (e.g., surveys, listening sessions, data reviews)?

- How will you adjust implementation based on student and counselor input?

Section 9: Monitoring, Evaluation & Adjustment (Ch. 8, Ch. 10)

- What indicators will you use to measure success?

How will you monitor for bias, access disparities, or unintended harm?

- What is your plan for continuous review and improvement?

Section 10: Advocacy, Scaling & Sharing (Ch. 10)

- How will lessons learned be documented and shared?

- What advocacy efforts will you lead or support at the district/state level?

- How will students and families continue to be partners in future AI innovation?

Signatures & Commitment

Lead Counselor/Coordinator:

Date:

Team Members:

Date:

District/School Administrator:

Date:

This template can be adapted for site-level, district, or regional planning efforts.

Appendix 3

Counselor's AI Evaluation Checklist

Use this checklist before adopting or approving any AI-driven tool for counseling or student services:

Purpose and Relevance

- Does the tool clearly align with a student-centered counseling goal?
- Is it addressing a need identified by students, counselors, or data?

Transparency and Consent

- Are students and families informed about how the tool works?
- Can students opt out or choose how their data is used?
- Is informed consent obtained in an accessible and age-appropriate way?

Data Use and Privacy

- What data does the tool collect, and how is it stored?

- Who has access to student data—now and in the future?
- Does it comply with FERPA, COPPA, HIPAA, and local privacy laws?

Equity and Cultural Responsiveness

- Was the tool tested across diverse racial, linguistic, gender, and neurodiverse populations?
- Can the tool be adapted for different cultural contexts or languages?
- Does it risk reinforcing stereotypes or disparities?

Human Oversight

- Is there a clear process for counselor review and interpretation of AI-generated outputs?
- What happens if the tool misflags or misguides a student?

Ongoing Feedback and Improvement

- Is there a process for collecting student and counselor feedback?
- Can the system be updated or improved based on feedback?

Appendix 4

Student Consent & Transparency Template for AI Tools

[School/District Name] is committed to protecting your privacy and ensuring you understand how technology is used to support your education and well-being.

This tool will:

- Help us identify how to support you academically, socially, or emotionally
- Analyze information you choose to share (such as journal entries or survey responses)
- Provide suggestions or alerts to your counselor if needed

This tool will NOT:

- Diagnose or treat mental health conditions
- Make decisions without human review
- Be used for discipline or academic penalties

You have the right to:

- Know how your data is being used
- Ask questions at any time
- Decline or opt out of using this tool
- Request to see your data or have it removed (where possible)

I understand how this tool works and consent to use it:

Signature: _____   Date: _____

Parent/Guardian Signature (if under 18): _____

Appendix 5
Equity Review Questions for Institutional Teams

Use these questions during the planning or evaluation phase of any AI implementation:

1. Who is most likely to benefit from this tool? Who might be left out?

2. Was the tool developed with input from diverse students, educators, and families?

3. Could this tool misrepresent students with disabilities, language differences, or trauma histories?

4. What checks are in place to identify and respond to bias?

5. How will we track usage patterns across race, gender, and socioeconomic groups?

6. What community partners should be consulted to ensure inclusivity?

Appendix 6
AI Tool Pilot Planning Guide

1. Define the goal

- What student need or counselor priority does this tool address?

2. Form a pilot team

- Include a counselor, teacher, administrator, student representative, IT lead, and parent/family liaison.

3. Select a scope

- Choose a manageable group (e.g., one grade, department, or school site)

4. Develop a feedback loop

- Weekly check-ins with pilot participants
- Surveys for students and staff
- Student voice focus groups

5. Monitor data and ethics

- Document flagged concerns, false positives, or student complaints
- Keep records of who accesses the data and how it's used

6. Assess and report

- Analyze results by subgroup (e.g., race, language, gender)
- Prepare a short report with findings, challenges, and recommendations

Endnotes

1. HolonIQ, 2025 Global Education Outlook, November 25, 2024, https://www.holoniq.com/notes/2025-global-education-outlook.

2. Dan Fitzpatrick, "76% of Teachers Feel Unprepared for AI. Why Training Is Urgent," Forbes, March 15, 2025, https://www.forbes.com/sites/danfitzpatrick/2025/03/15/76-of-teachers-feel-unprepared-for-ai-why-training-is-urgent/.

3. American School Counselor Association, "School Counselor Roles & Ratios," accessed April 29, 2025, https://www.schoolcounselor.org/about-school-counseling/school-counselor-roles-ratios.

4. Lauraine Langreo, "More Teachers Say They're Using AI in Their Lessons. Here's How," Education Week, March 6, 2025, https://www.edweek.org/technology/more-teachers-say-theyre-using-ai-in-their-lessons-heres-how/2025/03.

5. Hannah Mayer et al., "Superagency in the Workplace: Empowering People to Unlock AI's Full Potential," McKinsey & Company, January 28, 2025, https://www.mckinsey.com/capabilities/mckinsey-digital/our-insights/superagency-in-the-workplace-empowering-people-to-unlock-ais-full-potential-at-work.

6   American School Counselor Association, "The School Counselor and Student Safety and Digital Technology," American School Counselor Association, accessed April 29, 2025, https://www.schoolcounselor.org/Standards-Positions/Position-Statements/ASCA-Position-Statements/The-School-Counselor-and-Student-Safety-Digital.

7   Centers for Disease Control and Prevention, "Youth Risk Behavior Survey Data Summary & Trends Report: 2013–2023" (U.S. Department of Health and Human Services, 2024), https://www.cdc.gov/yrbs/dstr/index.html.

8   Center for Community College Student Engagement, "Supporting Minds, Supporting Learners: Addressing Student Mental Health to Advance Academic Success" (Austin, TX: University of Texas at Austin, 2024), https://www.ccsse.org/reports/Supporting-Minds.pdf.

9   U.S. Department of Education, Office of Educational Technology, "Artificial Intelligence and the Future of Teaching and Learning: Insights and Recommendations" (Washington, DC: U.S. Department of Education, 2023), https://www.ed.gov/sites/ed/files/documents/ai-report/ai-report.pdf.

10 Elizabeth Laird, Maddy Dwyer, and Hugh Grant-Chapman, "Off Task: EdTech Threats to Student Privacy and Equity in the Age of AI" (Washington, DC: Center for Democracy & Technology, September 20, 2023), https://cdt.org/wp-content/uploads/2023/09/091923-CDT-Off-Task-web.pdf.

11 Julia H. Kaufman et al., "Uneven Adoption of Artificial Intelligence Tools Among U.S. Teachers and Principals in the 2023-2024 School Year," February 11, 2025 (Santa Monica, CA: RAND Corporation, 2025), https://www.rand.org/pubs/research_reports/RRA134-25.html.

12 American Counseling Association, "AI Can Support—But Not Replace—Human Counselors, According to New Recommendations," American Counseling Association, January 25, 2024, https://www.counseling.org/publications/media-center/article/2024/01/25/ai-can-support-not-replace-human-counselors-according-to-new-recommendations.

13 American Counseling Association, "Integrating AI and LLMs into Counseling Education: Ethical and Inclusive Approaches," accessed April 29, 2025, https://www.counseling.org/resources/research-reports/artificial-intelligence-counseling/for-faculty.

14 Obed Boateng and Bright Boateng, "Algorithmic Bias in Educational Systems: Examining the Impact of AI-Driven Decision Making in Modern Education," *World Journal of Advanced Research and Reviews* 25, no. 1 (2025): 2012-2017, https://doi.org/10.30574/wjarr.2025.25.1.0253.

15 U.S. Department of Education, "Artificial Intelligence and the Future of Teaching and Learning."

16 Center for Democracy & Technology, "Out of Step: Students, Teachers in Stride with EdTech Threats While Parents Are Left Behind," January 2025, https://cdt.org/wp-content/uploads/2024/12/FINAL-SLIDES-CDT-Polling-2024-Supporting-Slides-112624-FINAL-CF-MM.pdf.

17 Jake Bryant, Christine Heitz, Saurabh Sanghvi, and Dilip Wagle, "How Artificial Intelligence Will Impact K–12 Teachers," McKinsey & Company, January 14, 2020, https://www.mckinsey.com/industries/education/our-insights/how-artificial-intelligence-will-impact-k-12-teachers.

18 Cecilia Ka Yuk Chan and Louisa H.Y. Tsi, "Will Generative AI Replace Teachers in Higher Education? A Study of Teacher and Student Perceptions," Studies in Educational Evaluation 83 (2024): 101395, https://doi.org/10.1016/j.stueduc.2024.101395.

19 Laird, Dwyer, and Grant-Chapman, "Off Task."

20 Digital Education Council, What Students Want: Key Results from DEC Global AI Student Survey 2024, August 7, 2024, https://www.digitaleducationcouncil.com/post/what-students-want-key-results-from-dec-global-ai-student-survey-2024.

www.ingramcontent.com/pod-product-compliance
Lightning Source LLC
Chambersburg PA
CBHW060353110426
42743CB00036B/2972